Trilobites
Common Trilobites of North America

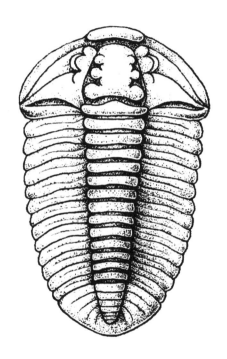

Text and Illustrations by Jasper Burns

A NatureGuide Book

Published by NatureGuide Books, an imprint of Miller's Fossils, Inc. 16 Marsh Woods Lane, Wilmington, DE 19810 (302-475-8819)

Cover and book design by Donald S. Miller

Publisher's Cataloging-in-Publication Data

Burns, Jasper.
     Trilobites : common trilobites of North America / text and illustrations by Jasper Burns. -- 1st ed.
     p. cm
     Includes bibliographical references and index.
     LCCN: 98-89608
     ISBN: 0-9669157-0-4

     1. Trilobites--North America--Popular works.
2. Trilobites--North America--Identification.
I. Title.

QE821.B87 1999   565.39'097
                         QBI98-990006

CONTENTS

INTRODUCTION

Trilobites are among the most interesting and scientifically important of all prehistoric animals. They are the first large, complex creatures to appear in the fossil record, dating back to the beginning of the Cambrian period, 570 million years ago. These earliest known forms were already highly developed with a long evolutionary history behind them. Trilobites flourished throughout the Paleozoic Era from 570 – 245 million years ago, evolving into a bewildering variety of forms (more than 1,500 genera and 12,000 species are known so far), before vanishing at the end of the Permian period. They had inhabited the Earth for at least 325 million years, more than twice as long as the dinosaurs.

As fossils, trilobites possess an almost unrivalled ability to fascinate. Their symmetrical forms and elegant simplicity – and occasionally baroque ornamentation – combine with the pleasing rhythms of their segmentation to captivate and delight the eye. They are probably the most highly prized of invertebrate fossils and are eagerly collected around the world.

DESCRIPTION

All trilobites are variations on a simple theme. They are arthropods (as are insects, barnacles, crabs, and spiders), with hard, external skeletons and jointed appendages. All known species were marine, and all were protected by carapaces (or tests) that were divided into three sections from back to front and from side to side. The back to front divisions are familiar in the animal kingdom – head (or cephalon), thorax (always segmented), and tail (or pygidium). The side to side division into three lobes with a central axis and two lateral pleural lobes is distinctive of the group and gives trilobites their name (from the Greek *trilobos* or "three lobed"). Some trilobites, such as *Isotelus*, *Bumastus*, and *Trimerus*, exhibit an indistinct division into three lobes although they are still evident, especially in younger individuals. Each thoracic segment of a trilobite consists of two lateral pleura (singular: pleuron) and a central, axial ring to which they are attached.

As adults, trilobites ranged from less than ¼ inch to 30 inches in length, but the vast majority of species measured between ¾ inch and 3 inches. Some forms were literally as flat as a pancake while others were almost cylindrical in shape. Carapaces could be nearly circular or very long and slender, though most were moderately elongated ovals.

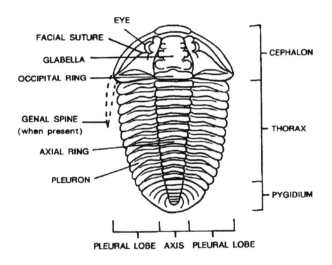

Figure 1

LIFESTYLE
The manner of life followed by various trilobites is open to question and often highly controversial. It is taken for granted that most species were scavengers who crawled along the sea floor. Some were apparently adapted for burrowing and others may have been active swimmers. Trilobites lacked jaws, teeth, claws, or mandibles and were presumably incapable of overpowering and eating anything more formidable than a small worm. Some species were probably herbivorous, feeding on algae or seaweed, while others may have ingested sediment rich in organic material, as earthworms do on land today.

While it is doubtful that trilobites were active predators, it is obvious that they were often potential prey. Their heavily armored carapaces frequently bear spines and bosses, and most species, especially post-Cambrian forms, were capable of enrolling themselves into a ball, protecting their soft undersides from their enemies. A highly developed ability to detect and avoid danger is also suggested by trilobite eyes and other sensory structures.

Enrolled trilobite
(*Flexicalymene meeki* Foerste)
Figure 2

EYES
Trilobite eyes are the first to appear in the fossil record and could be extremely complex. Like those of modern insects, they were compound, with as few as 14 facets per eye to as many as 15,000 facets per eye in some species! Two main types of trilobite eyes are recognized: holochroal eyes, the most common type, and schizochroal eyes, which only occur in members of two families of the Order Phacopida.

Schizochroal eye   Holochroal eye
Phacops   *Bumastus*
Figure 3

Holochroal eyes have numerous lenses which are in contact with each other and are all covered by a single membrane. The first and last trilobites known had this kind of eye, which is similar in structure to the eyes of living arthropods. Schizochroal eyes are unique in the animal kingdom. The individual lenses, much larger than in holochroal eyes, do not touch each other and each has its own corneal membrane. These eyes conferred excellent, probably stereoscopic vision.

Some trilobites, such as the Agnostids, seem to have been totally blind. But on the whole, trilobite vision must have been quite good. As they were probably not predators of large animals, their excellent vision was primarily a defensive adaptation. Large, protruding eyes may have been suited for night vision or turgid waters, or for living in highly complex and dangerous shallow water habitats. They would not have been convenient for burrowing – small eyes set close to the carapace would serve much better for this way of life. Blind trilobites may have been burrowers , or inhabitants of dark, deep water habitats, or may have relied on other senses for protection.

GROWTH
Like all arthropods, trilobites had to molt their exoskeletons periodically in order to grow. Immediately after molting, their new skeletons would have been very soft, like a soft-shell crab's, and fossils of trilobites in this condition have been found. It has been estimated that a typical trilobite would molt 20 or more times during its lifetime. The cast-off bits of exoskeleton were hard and could be preserved as fossils as easily as the remains of a dead animal. The majority of trilobite fossils are the products of molting.

In order to crawl out of the old, hard skins, most trilobites were capable of disassembling their heads along lines known as facial sutures. The position of these sutures varied from species to species and was used to classify trilobites until it was discovered that the same arrangement could evolve separately in distantly related forms. The term "free cheeks" is used to describe the parts of the cephalon which separated from the "cranidium," or the central glabella, and the area of the cephalon that remained attached to it.

As trilobites grew and molted, they also changed in form, going through a series of growth stages in a sort of metamorphosis. Typically, the number of thoracic segments increased with development while the shape of the cephalon and pygidium often showed considerable change. A larval trilobite, known as a protaspid, gradually grew into an intermediate stage, called a meraspid, which finally matured into an adult holaspid.

LOCOMOTION AND FEEDING
While trilobite appendages are rarely preserved, a few spectacular exceptions to this rule have revealed many details about their structure. A trilobite leg was biramous, meaning that it branched into two jointed structures: one for walking and one bearing a gill for breathing. Each thoracic segment bore one pair of legs and the cephalon and pygidium (which were formed from the coalescence of several body segments) each had a variable number of pairs. A pair of antennae projected from the front of the animal and some species also possessed a pair of antenna-like cerci at the tail end.

Trilobite trackway
Figure 4

Food was gathered and chewed by the action of spines on the inner parts of the legs as the trilobite moved. The trilobite mouth was very small and opened into the stomach that was positioned directly below the central swelling on the trilobite's head (glabella). An enlarged glabella is interpreted as a sign that a species had a large stomach, perhaps for processing low-grade food such as sediment. Directly beneath the glabella was the hypostome, a hard structure detached from the rest of the skeleton. This often rested against the ventral, flange-like extension of the cephalon (doublure). The mouth opening was directly behind the hypostome, which presumably allowed a structural flexibility that was important in feeding.

Trilobite legs were much the same in structure and function across species. This has been cited as one reason why trilobites gradually became less numerous as time passed; they may have been unable to compete with more specialized and sophisticated arthropods in which the legs developed into a wide variety of structures with many functions (e.g. claws). However, it is a fact that trilobites outlived some of their more complex relatives who had differentiated limbs, such as the Burgess Shale trilobitoids and the formidable eurypterids. Any body plan that persisted for 325 million years and survived the rise of the fishes, nautiloids, and other efficient predators must have been highly adaptive.

EXTINCTION

The demise of the trilobites is attributed to the severe reduction in shallow water marine habitats that resulted from the forming of the super-continent, Pangea, at the end of the Paleozoic Era. When the Earth's land area is distributed in many separate continents, as it is today, nearshore environments are at a maximum, but one vast landmass is surrounded by much less shallow water for trilobites and other marine animals to live in. This geologic event coupled with more sophisticated competitors and predators all but assured extinction. The closest living relatives of the trilobites are found among the horseshoe crabs and crustaceans.

CLASSIFICATION

There have been numerous schemes for trilobite classification through the years, and the matter is probably not finally settled. This guide recognizes eight orders of the Class Trilobita as follows: Redlichiida, Corynexochida, Agnostida (all three of these orders were extinct by the end of the Ordovician), Odontopleurida, Phacopida, Ptychopariida, Lichida (these four orders disappeared by the end of the Devonian), and Proetida (which persisted until the end of the Permian). Trilobites were very common during the Cambrian, but their numbers and diversity steadily decreased afterwards. They were not common after the Devonian, and near their final extinction at the end of the Permian, only one family (Phillipsiidae) was still in existence.

ABOUT THIS BOOK

The species illustrated on the following pages were selected to show the diversity of trilobites as well as to introduce some of the more commonly collected forms. With a couple of exceptions, these species are North American in origin, though nearly all of them have close relatives in other parts of the world. European and North African trilobites have very close affinities with North American forms, reflecting the close geographic proximity of the areas when trilobites lived. New species are still being discovered throughout the world, especially in South America, Africa, Russia, and parts of Asia, and new techniques of preparation (such as air abrasive removal of matrix) has revealed many new details of trilobite anatomy. Trilobites have been around for a very long time, but their rediscovery by humans is still in its early stages.

A few comments about where trilobites are found: The names in parentheses after the localities given refer to the rock units (usually formations) in which the particular species of trilobite may be found. The lists of collecting areas for each species are not comprehensive but are partial lists of published occurrences. Many of the trilobite species may be found in other states or countries that were not mentioned in the sources consulted.

Any region which has exposures of Paleozoic sedimentary rocks, especially Cambrian through Devonian in age, is a likely source of trilobite fossils. In the U.S., Ohio, Nevada, Oklahoma, Utah, and New York are particularly well known for their wealth of unusual and well-preserved trilobite fossils, but many other states have a comparable abundance and diversity of specimens.

Suggested Reading

The Audobon Society Field Guide to Fossils, Thompson, Ida, Alfred A. Knopf, new York, 1982

The Fossil Book, Fenton, C. and M, Rich, P. and T., Doubleday, New York, 1989.

Simon and Schuster's Guide to Fossils, Arduini, P., and Teruzzi, G., Simon and Schuster, New York, 1986.

Trilobites, Levi-Setti, Riccardo, Second Edition, University of Chicago Press, Chicago and London, 1993.

Trilobites, Whittington, H. B., Fossils Illustrated Volume II, Boydell Press, Woodbridge, Suffolk, England, 1992.

Treatise on Invertebrate Paleontology: Part O – Arthropoda 1, Edited by Raymond C. Moore, University of Kansas Press, Lawrence, Kansas, 1959.

Trilobites of the Thomas T. Johnson Collection, Johnson, Thomas T., self-published, 1985.

Callavia bröggeri Walcott

Geologic range: Lower Cambrian

Order: Redlichiida

Family: Callaviidae

Where found: Newfoundland (Brigus)

COMMENTS:
This species has a tiny pygidium and the last two of the eighteen thoracic segments are reduced in size. Related species may be found in Europe. The glabella has a long, posteriorly directed spine, which distinguishes it from the similar North American genus, *Holmia*. Specimens of *Callavia* may be as much as 6 inches (25 cm.) in length.

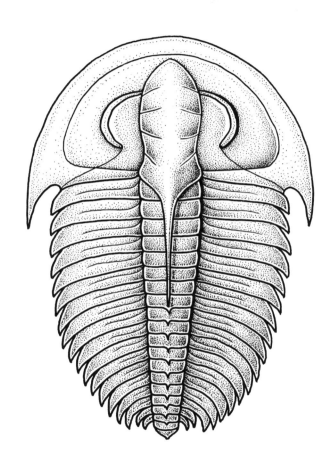

Olenellus clarki Resser

Geologic range: Lower Cambrian

Order: Redlichiida

Family: Olenellidae

Where found: California (Cararra), Nevada (Pioche)

COMMENTS:
Similar species in Vermont, Western Canada, the Appalachians.
Olenellus is one of the most primitive trilobites. It has a very
small pygidium and up to 44 or even more thoracic segments.
One of the thoracic segments bears a long spine. Some or all of
the segments posterior to this spine are usually missing from fossil
specimens. When present, their form is so similar to that of
segmented worms that they have been used to support a theory
that trilobites developed from worms at some time before the
Cambrian period. *Olenellus* is believed to have dug small pits on
the sea floor in search of food. This species attained 3 ½ inches (8
cm.) in length.

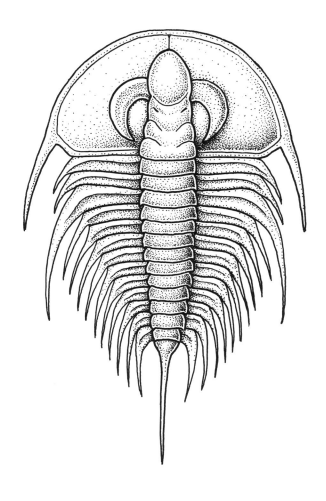

Albertella helena Walcott

Geologic range: Middle Cambrian

Order: Corynexochida

Family: Zacanthoididae

Where found: Montana (Gordon), British Columbia, Nevada

COMMENTS:
Similar to *Zacanthoides* but with smaller eyes, narrower brim, and wider fixed cheeks. Some scientists have speculated that the enlarged thoracic pleurae (macropleurae) on this and other kinds of trilobites were reproductive structures from which eggs or sperm were released through small holes. This species may exceed 2 ½ inches (6 cm.) in length.

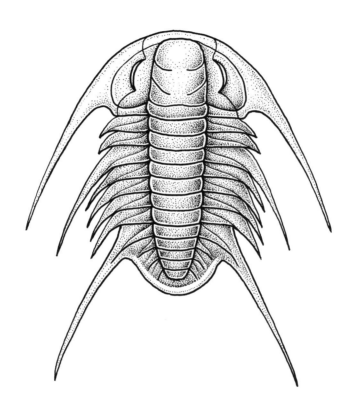

Alokistocare harrisi Robison

Geologic range: Middle Cambrian

Order: Redlichiida

Family: Alokistocaridae

Where found: Utah (Wheeler)

COMMENTS:
Trilobites of the order Redlichiida are considered the most primitive members of the Class. Characteristically, they have extremely small pygidia and numerous thoracic segments.

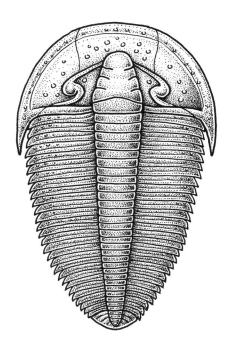

Elrathia kingii Meek

Geologic range: Middle Cambrian

Order: Ptychopariida

Family: Alokistocaridae

Where found: Utah (Wheeler)

COMMENTS:
Well-preserved examples with shiny black exoskeletons from the Wheeler Amphitheater in Millard County, Utah are perhaps the most frequently purchased American trilobite. Complete specimens range in size from 1/8 inch (3 mm.) to 2 inches (5 cm.). The related species *E. georgiensis* occurs in Georgia.

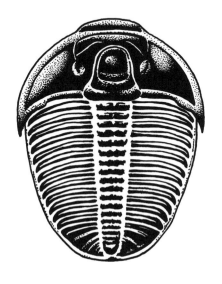

Ogygopsis klotzi Rominger

Geologic range: Middle Cambrian

Order: Corynexochida

Family: Ogygopsidae

Where found: British Columbia (Stephen)

COMMENTS:
This trilobite has small, widely spaced eyes and occurs in the famous Burgess Shale. Specimens may attain 3 ½ inches (8 cm.) in length.

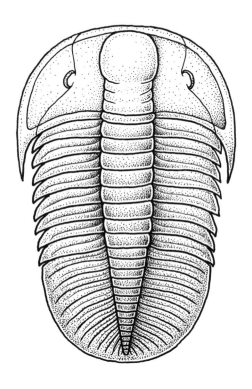

Olenoides serratus Rominger

Geologic range: Middle Cambrian

Order: Corynexochida

Family: Dorypygidae

Where found: British Columbia (Stephen)

COMMENTS:
Similar to *Kootenia* but pygidium has both grooves and furrows rather than grooves only. This species, like *Ogygopsis*, is part of the famous Burgess Shale fauna. Sometimes, specimens are found which show incredible details of the appendages. *Olenoides* had a pair of long dorsal antennae followed by 3 pairs of legs under the head, 7 pairs of legs under the 7 thoracic segments, and 4 to 6 pairs of legs under the pygidium. At the posterior end of the trilobite was a pair of antenna-like cerci. As in other trilobites, all of the legs were similar in design. Each leg was branched into two jointed structures - one for walking and one with gills for breathing. The legs under the head were also designed to enable the trilobite to "chew" food as it walked, but there were no claws or mandibles or teeth. Still, *Olenoides serratus* in considered to have been capable of seizing and eating small worms. Related species of *Olenoides* are found in Alabama and Utah.

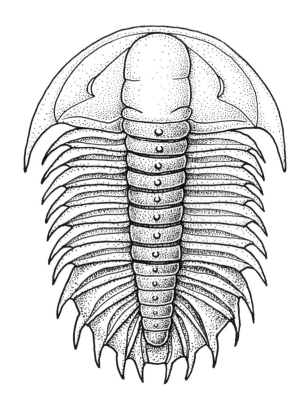

Orria elegans Walcott

Geologic range: Middle Cambrian

Order: Corynexoichida

Family: Dolichometopidae

Where found: Utah (Marjum)

COMMENTS:
Similar to *Ogygopsis* but has narrower cranidium and the pygidium is longer and with a shorter axial lobe. *Orria* is a widespread genus in western North America.

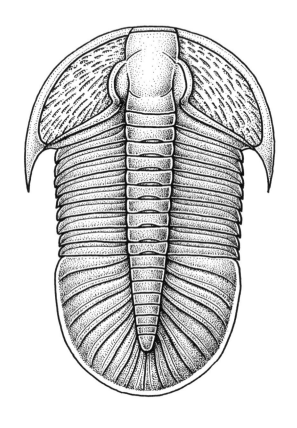

Paradoxides gracilis Boeck

Geologic range: Middle Cambrian

Order: Redlichiida

Family: Paradoxididae

Where found: Bohemia (Jinetz)

COMMENTS:
Species of *Paradoxides* occur in Massachusetts, Newfoundland, New Brunswick, Europe, northern Africa, Asia, and possibly Australia. This large Cambrian trilobite (to 20 inches (50 cm.)) is found in shallow water mudstones, shales, and sandstones. Its name derives from its surprising occurrence in easternmost North America, though it is so different from other North American trilobites and so similar to many Old World forms. The "paradox" was solved with the acceptance of continental drift and the realization that the *Paradoxides*-bearing sediments in North America had once been part of Europe.

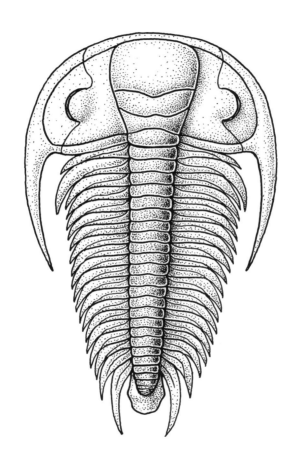

Peronopsis interstricta White

Geologic range: Middle Cambrian

Order: Agnostida

Family: Spinagnostidae

Where found: Utah (Wheeler)

COMMENTS:
An example of an order of small trilobites in which the thorax is reduced to only 2 or 3 segments and the cephalon and pygidium are approximately equal in size. Like all agnostids, *Peronopsis* was quite small, usually between ¼ and 1/3 inches (6-8 mm.) in length. These trilobites have no eyes on the dorsal surface and were probably blind. Speculation about their lifestyle has run the gamut. They were believed to have been planktonic, carried widely by ocean currents. Their flattened shape led some to suggest they were ectoparasites who attached themselves to free-swimming hosts. However, their preservation in heaps of organic debris suggests the most likely solution: they were burrowers who spent their lives scavenging for food. Other species of *Peronopsis* occur in Montana, Siberia, and Europe.

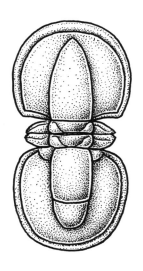

Zacanthoides typicalis Walcott

Geologic range: Middle Cambrian

Order: Corynexochida

Family: Zacanthoididae

Where found: Nevada (Chisholm)

COMMENTS:
This genus is widespread throughout North America. Its numerous spines presumably stabilized it in soft sediments while providing protection from enemies.

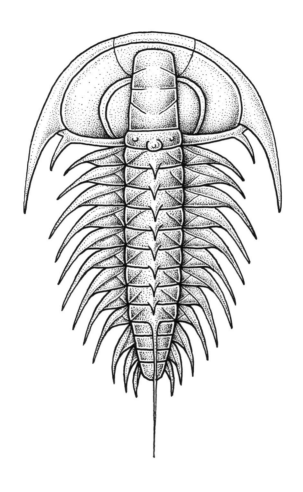

Cedaria minor Walcott

Geologic range: Upper Cambrian

Order: Ptychopariida

Family: Raymondinidae

Where found: Utah (Weeks)

COMMENTS:
Free cheeks are strongly attached to the cranidium so that heads
are often found complete.

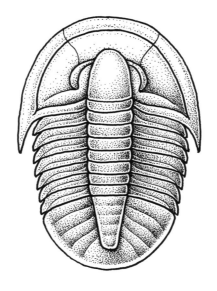

Crepicephalus iowensis Owen

Geologic range: Upper Cambrian

Order: Ptychopariida

Family: Crepicephalidae

Where found: Iowa (Dresbachian), Upper Mississippi Valley

COMMENTS:
Thorax has 12 to 14 segments. Related species may be found from Texas to Alabama and in Wyoming and Utah. Some members of this genus exceeded 6 inches (15 cm.) in length.

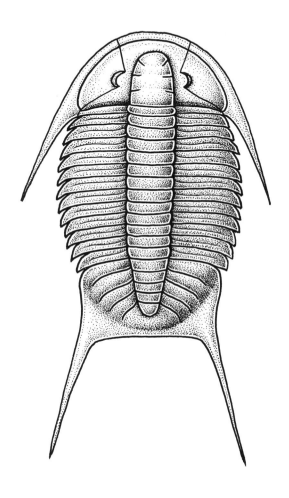

Dikelocephalus oweni Ulrich and Resser

Geologic range: Upper Cambrian (Trempealeau)

Order: Ptychopariida

Family: Dikelocephalinidae

Where found: Wisconsin, Upper Mississippi Valley

COMMENTS:
A widespread genus, but not common anywhere. Related species occur in Minnesota and in Europe. The pygidium in Dikelocephalus may approach 4 inches (10 cm.) in width.

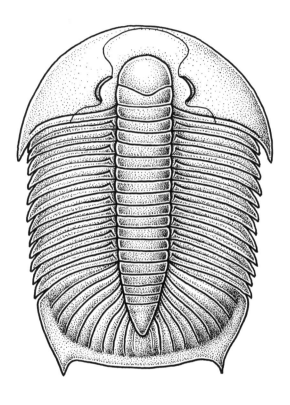

Ampyxina scarabeus Butts

Geologic range: Ordovician

Order: Ptychopariida

Family: Raphiophoridae

Where found: Virginia (Liberty Hall)

COMMENTS:

This trilobite had no eyes. It often occurs in the same strata as *Dionide holdeni*. A similar species, *A. bellatula*, is found in Missouri.

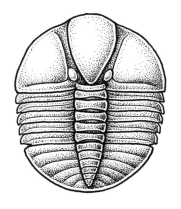

Ceraurus pleurexanthemus Green

Geologic range: Ordovician (Black River, Trenton)

Order: Phacopida

Family: Cheiruridae

Where found: Widespread throughout eastern and central North America (e.g. Ontario, Virginia, Pennsylvania).

COMMENTS:
Similar to *Cheirurus* but *Ceraurus* has 2 tail spines as opposed to 7. Related species occur in Europe. Complete specimens of *Ceraurus* are rare but isolated pygidia are fairly common. The appendages of this genus are better known than for most trilobites and are distinctive in that the gill branches are modified into paddle-shaped structures that may have been useful in swimming. Specimens may exceed 2 inches (5 cm.) in length.

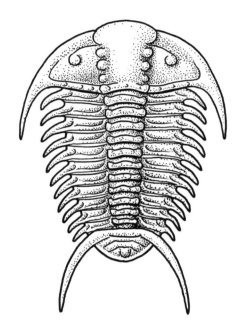

Cryptolithus tesselatus Green

Geologic range: Ordovician (Trenton)

Order: Ptychopariida

Family: Trinucleidae

Where found: Alabama, Kentucky (Latonia), Maryland, Pennsylvania, Virginia, West Virginia (Martinsburg), Quebec, and Ontario

COMMENTS:
Similar species are found in Europe. This eyeless trilobite has a distinctive perforated fringe on the cephalon. There has been much conjecture about the purpose of this fringe. It may have been used as a filter to capture tiny prey or served like a snowshoe to keep the animal from sinking into soft bottoms. Another theory is that the holes contained sensors which helped the trilobite orient itself to underwater currents. *Cryptolithus* was not designed for swimming and probably excavated shallow burrows on the sea floor, with its head pointing into the current. An inability to adjust to shifting currents might have exposed the animal to being overturned or swept away. The swollen glabella on this trilobite may suggest that it had a large stomach and ingested quantities of sediment to extract the organic nutrients. The long genal spines on this species account for more than half of the 1 ½ inch (4 cm.) maximum length. Occasionally found enrolled.

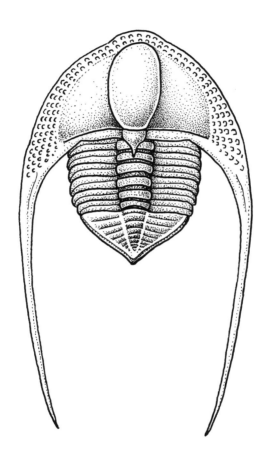

Dionide holdeni Raymond

Geologic range: Ordovician

Order: Ptychopariida

Family: Dionididae

Where found: Virginia (Liberty Hall)

COMMENTS:
In Virginia, this species occurs with *Ampyxina scarabeus*.
Related species may be found in Europe. The long genal spines,
perforated margin of the cephalon, and relatively large glabella
may suggest some parallels with *Cryptolithus* regarding this
species' way of life.

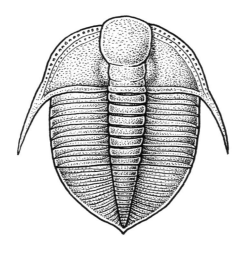

Triarthrus eatoni Hall

Geologic range: Ordovician

Order: Ptychopariida

Family: Olenidae

Where found: New York (Utica), Pennsylvania (Reedsville), West Virginia, Ontario

COMMENTS: Exceptionally well-preserved examples of this species from the so-called "Beecher's trilobite bed" (Frankfort Shale, Rome, New York) are famous for showing details of appendages that are preserved in pyrite. These features can be seen by x-raying rocks that contain specimens (pyrite is not transparent to x-rays) , or through special preparation and photographic techniques. Related species occur in South America and in Europe.

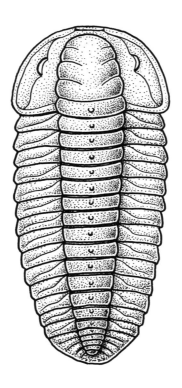

Basiliella barrandei Hall

Geologic range: Middle Ordovician

Order: Ptychopariida

Family: Asaphidae

Where found: Minnesota, New York (Trenton), Virginia, Wisconsin, Ontario

COMMENTS:
Trilobites of this genus have been found in Argentina. The huge cephalon of this species, reminiscent of the cephalothorax of a horseshoe crab, must have provided protection as well as stability.

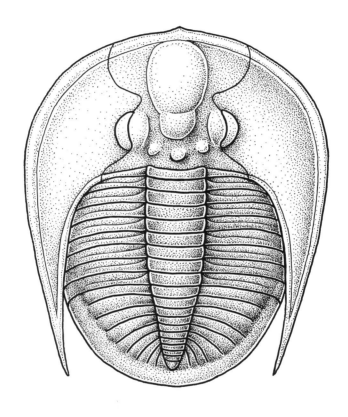

Isotelus gigas Dekay

Geologic range: Middle Ordovician

Order: Ptychopariida

Family: Asaphidae

Where found: New York (Trenton), widespread through North America

Comments:

Related species in Norway. Similar to *Homotelus*, but unlike that genus, *Isotelus* has concave brims on the cephalon and pygidium. Because of its smooth carapace and broad, flat margins, this trilobite is believed to have been capable of burrowing. *Isotelus gigas* attains 2 inches (5 cm.) in length but related species were much larger. The state fossil of Ohio, *I. maximus*, exceeded 12 inches (30 cm.) and specimens of another species of *Isotelus* as much as 30 inches (75 cm.) long have been found, ranking among the largest of all known trilobites.

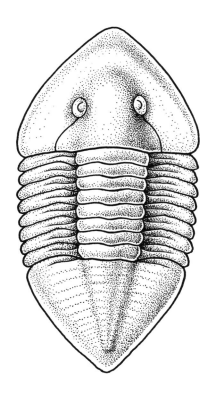

Scutellum lunatum Billings

Geologic range: Middle Ordovician (Trenton)

Order: Ptychopariida

Family: Thysanopeltidae

Where found: New Jersey, Minnesota, Ontario, Manitoba

COMMENTS:
This genus, which survived from the Ordovician to Devonian periods, is rare in North America but common in Europe.

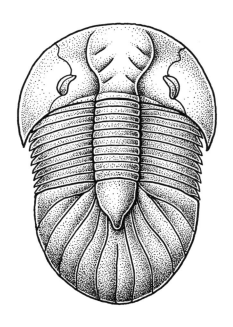

Flexicalymene meeki Foerste

Geologic range: Upper Ordovician

Order: Phacopida

Family: Calymenidae

Where found: Kentucky (McMillan), Ohio (Waynesville), New York, Pennsylvania, Virginia, West Virginia (Martinsburg), and through much of central North America

COMMENTS:
Very similar to the Silurian to Middle Devonian genus *Calymene*, but the leading edge of the cephalon is more sharply upturned. Some experts suggest that this species was a burrower – the sturdy, plow-like cephalon seems well-adapted for rummaging through debris on the ocean floor. Specimens may approach 4 inches (10 cm.) in length, but most are closer to 1 inch (2.5 cm.).

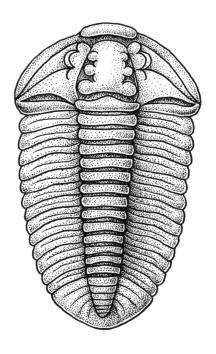

Arctinurus boltoni Bigsby

Geologic range: Silurian (Clinton)

Order: Lichida

Family: Lichidae

Where found: New York, Ontario

COMMENTS:
Pygidium wider than long. Specimens of *Arctinurus* are usually found with the eyes crushed. The front of the cephalon forms a snout-like projection, suggesting that *Arctinurus* plowed through debris on the sea floor in search of food. Specimens may exceed 5 inches (12 cm.) in length.

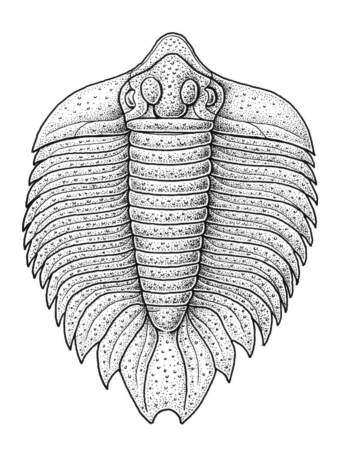

Bumastus niagarensis Whitfield

Geologic range: Silurian (Niagaran)

Order: Ptychopariida

Family: Illaenidae

Where found: Illinois, Missouri, Ohio, Wisconsin

COMMENTS:
Similar species occur in Britain. Like other Illaenids, *Bumastus*
had a smooth carapace, which may suggest that it was a burrower.
Specimens generally occur in limestones, including coral reef
deposits. The holochroal eyes project well above the carapace. It
has been speculated that *Bumastus* and similar trilobites may have
rested their heads on the sea floor, with their eyes protruding
through a shallow layer of sediment. The rest of their bodies may
have been more deeply buried and oriented somewhat vertically.
Bumastus specimens may be up to 4 inches (16 cm.) long.

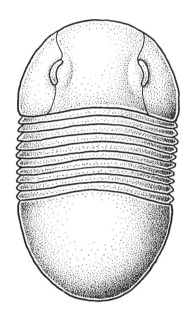

Calymene celebra Raymond

Geologic range: Silurian

Order: Phacopida

Family: Calymenidae

Where found: widespread in North America (e.g. Wisconsin (Niagaran), Illinois)

COMMENTS:
The state fossil of Wisconsin. This genus persisted for 75 million years, from the Lower Silurian to the Middle Devonian, and may be found in Europe, Australia, and North and South America. As with the related *Flexicalymene meeki*, specimens of enrolled individuals are common.

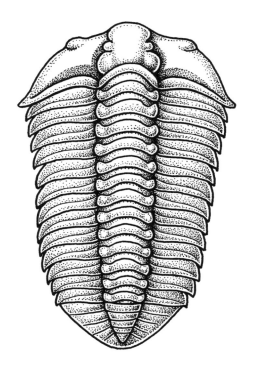

Dalmanites verrucosus Hall

Geologic range: Silurian (Niagaran)

Order: Phacopida

Family: Dalmanitidae

Where found: Arkansas (St. Claire), Illinois, Indiana (Waldron), Missouri (Bainbridge), Wisconsin

COMMENTS:
The genus *Dalmanites* is represented in rocks from Ordovician to Devonian age and may be found in Europe as well as North America. *Odontochile* is very similar, but its pygidium usually has more annulations (16 to 22) than *Dalmanites* (11 to 16). Like *Coronura aspectans*, this trilobite had large, schizochroal eyes that protruded abruptly from a broad, flat carapace. The long tail spine has been compared to the telson of horseshoe crabs, which the living animals use to maneuver in soft mud or to right themselves when overturned. This species attained lengths of 6 inches (15 cm.) but some Dalmanitids grew to more than one foot (30 cm.) long.

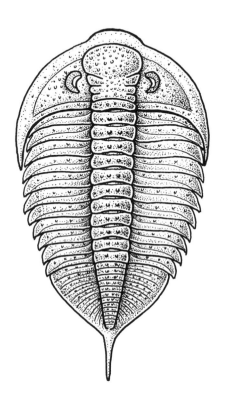

Deiphon forbesi Barrande

Geologic range: Silurian

Order: Phacopida

Family: Cheiruridae

Where found: Bohemia, England

COMMENTS:
Related species in China. Long-spined trilobites were once thought to have been active swimmers or drifters on ocean currents as their spines were believed to aid in buoyancy. However, it was realized that long spines would only assist very small animals in keeping afloat – even the one inch (2.5 cm.) long *Deiphon* was too large to benefit from this effect. More probably, this species was a bottom dweller whose spines assisted in staying on top of soft sediments. This view is supported by speculation that the large glabella housed a stomach used to process large quantities of sediment for their organic content.

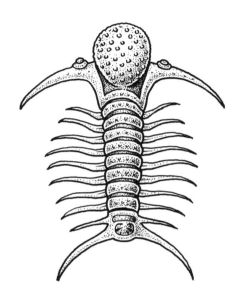

Encrinurus ornatus Hall and Whitfield

Geologic range: Middle Silurian (Niagaran)

Order: Phacopida

Family: Encrinuridae

Where found: New York (Niagaran), widespread through eastern and central North America

COMMENTS
Twenty incomplete rings on the axial lobe of the pygidium; a boldly ornate cephalon with prominent, stalked eyes; related species are found in Europe. The large bumps on the head of this trilobite may have contained sensors – similar bumps on trilobites have been shown to have a hollow space under them or canals connected to the underside of the exoskeleton. *Encrinurus* is common in shallow water deposits and its distinctive stalked eyes suggest an animal that kept a close watch on its surroundings. Nevertheless, this common Silurian type did not survive into the Devonian period. Two inches (5 cm.) is a good size for this species. Enrolled specimens are commonly found.

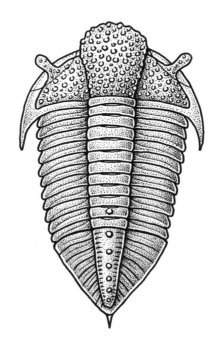

Trimerus delphinocephalus Green

Geologic range: Middle Silurian (Upper Clinton)

Order: Phacopida

Family: Homalonotidae

Where found: Maryland, New York (Rochester), Pennsylvania, Indiana, Kentucky, West Virginia, Ontario

COMMENTS:
Similar to *T. dekayi* but with much better defined segmentation on the pygidium and without a pitted carapace. This heavy, cylindrical species attained lengths of over 12 inches (30 cm.) and 10 inch (25 cm.) individuals are not uncommon.

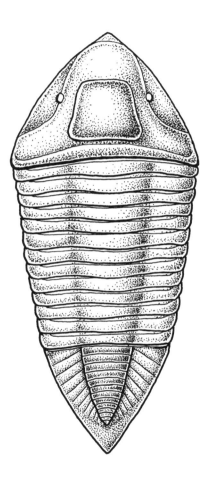

Basidechenella rowi Green

Geologic range: Devonian

Order: Proetida

Family: Proetidae

Where found: New York (Ludlowville), West Virginia (Mahantango)

COMMENTS:
Related species may be found in Europe. In contrast to some of the more specialized and spectacular-looking Devonian trilobites, *Basidechenella* was conservative in design, reminiscent of many Cambrian species (compare to *Cedaria major*). And yet, the only trilobites to survive into the succeeding geologic periods were members of the Order Proetida, similar in form to *Basidechenella* (see *Ameura, Griffithides, Breviphillipsia*). Specimens usually range between 1 and 2 inches (2.5-5 cm.) in length.

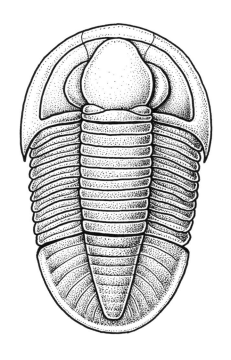

Odontopleura callicera Hall and Clarke

Geologic range: Devonian

Order: Odontopleurida

Family: Odontopleuridae

Where found: Ontario, Pennsylvania, and New York (Onondaga), Virginia and West Virginia (Needmore)

COMMENTS:
The entire margin of these ornate trilobites bristles with spines. Those on the margin of the cephalon were directed vertically downward in life and are believed to have held the head just above the sea floor while feeding. The spines of trilobites like *Odontopleura* are believed to have stabilized the animal and prevented it from sinking into soft substrates. They also may have had a protective function – some scientists have even speculated that the spines contained poison to ward off attackers.

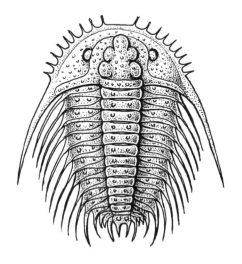

Phacops rana Green

Geologic range: Devonian

Order: Phacopida

Family: Phacopidae

Where found: Ontario, New York (Moscow), Pennsylvania (Hamilton), Maryland, Virginia, and West Virginia (Mahantango), Ohio (Silica), widespread in the central U.S.

COMMENTS: The state fossil of Pennsylvania is characterized by its extremely wide and swollen glabella and large, schizochroal eyes (see discussion under *Coronura aspectans*). Perhaps the best known and most frequently collected North American trilobite. Enrolled specimens are common, as are isolated heads and headless bodies. Most post-Cambrian trilobites were capable of enrollment, and it is assumed to have been a defensive posture. Studies of modern marine isopods suggest that it may also have had a function in feeding. Some isopods (which resemble trilobites somewhat and are among their closest living relatives) engage in filterfeeding while enrolled. Related species occur in Silurian as well as Devonian rocks in Europe and North Africa (e.g. Morocco where specimens of *P. megalonamicus* commonly exceed 6 inches (15 cm.) in length). *P. cristata* (Virginia, West Virginia) is similar but has pronounced axial spines. The related genus, *Reedops*, has smaller eyes and occurs in Oklahoma and Tennessee. Three inches (8 cm.) is a very good size for *Phacops rana*.

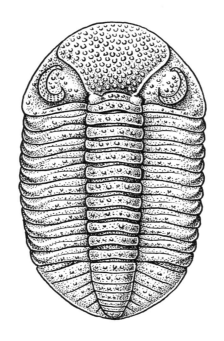

Trimerus dekayi Green

Geologic range: Devonian

Order: Phacopida

Family: Homalonotidae

Where found: New York, Pennsylvania (Hamilton), Maryland, Virginia and West Virginia (Mahantango)

COMMENTS:

This trilobite is presumed to have been a bottom feeder that plowed through sediments with its upturned snout in search of food. Frequently ascribed to the genus *Dipleura* in older references, and sometimes currently to the subgenus *Trimerus* (*Dipleura*), this large trilobite often exceeds 8 inches (20 cm.) in length. The surface of the test is uniformly pitted, so that steinkerns (internal molds) are covered with what appear to be short spines. It seems likely that the pits were related to sensory organs, perhaps involving the presence of numerous small "hairs." If so, this species may have appeared as a woolly trilobite. The genus also has representatives in Europe.

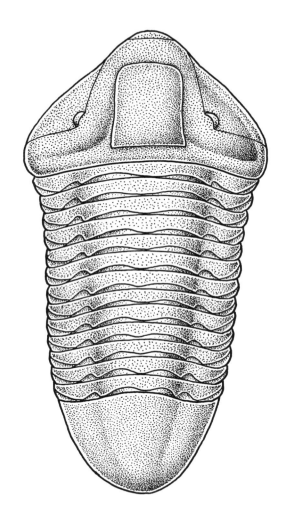

Coronura aspectans Conrad

Geologic range: Middle Devonian

Order: Phacopida

Family: Dalmanitidae

Where found: Widespread through Appalachians and Midwest (e.g. New York (Onondaga), Ohio, Indiana, West Virginia (Needmore)).

COMMENTS:
Quite large (5 inches (12 cm.) plus), ornate species; similar to *Synphoria* except for numerous spines on pygidium. Like other members of the Dalmanitid and the related Phacopid families, *Coronura* possessed large, multifaceted schizochroal eyes which gave it excellent, 360 degree vision and the ability to see in three dimensions. Schizochroal eyes (quite different in internal structure from the holochroal eyes found in other trilobites) were adapted for night vision and their lens structure featured a correction for astigmatism. The eyes of this species projected well above the broad, flat carapace and it is tempting to imagine that the trilobites covered themselves under a thin layer of sediment with their eyes exposed to watch for predators.

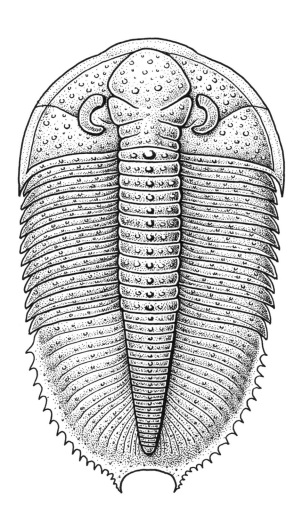

Odontocephalus aegeria Hall

Geologic range: Middle Devonian

Order: Phacopida

Family: Dalmanitidae

Where found: Pennsylvania, New York (Onondaga), Illinois and Missouri (Grand Tower), Ohio (Jeffersonville), Virginia, West Virginia (Needmore)

COMMENTS: The cephalic spines on this species have coalesced into a perforated fringe, and the pygidium ends in two spines, unlike the similar species *Synphoria. Odontocephalus* frequently exceeds 4 inches (10cm.) in length.   It has well-developed, schizochroal eyes.

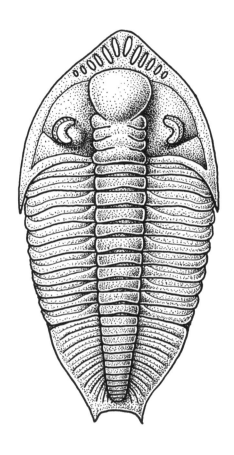

Terataspis grandis Hall

Geologic range: Middle Devonian

Order: Lichida

Family: Lichidae

Where found: New York (Onondaga), Ontario

COMMENTS:
Certainly one of the most spectacular of all trilobites, both in ornamentation and in size. Some specimens reach 27 inches (68 cm.) in length!

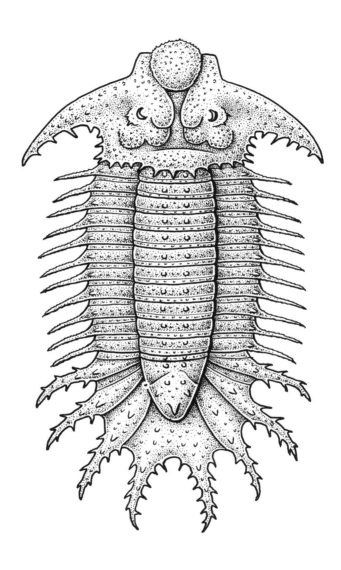

Greenops boothi Green

Geologic range: Middle and Upper Devonian

Order: Phacopida

Family: Dalmanitidae

Where found: New York (Moscow), Pennsylvania, Maryland, Virginia, and West Virginia (Mahantango), Ontario

COMMENTS:
Related forms are found in Europe, Africa, and the Middle East. This species, which may exceed 2 inches (5 cm.) in length, is found in shallow water limestones and shales. It has large schizochroal eyes like other Dalmanatids. The spines on this trilobite are blunt and very broad.

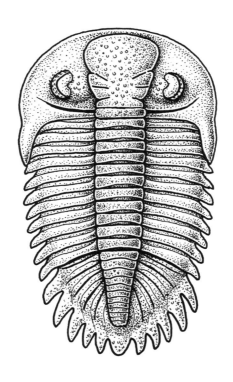

Breviphillipsia sampsoni Vogdes

Geologic range: Lower Mississippian

Order: Proetida

Family: Phillipsiidae

Where found: Missouri (Bushberg)

COMMENTS:
Common in shallow water limestones including reef deposits.
Closely related to *Griffithides*, these trilobites were formerly
ascribed to the genus *Phillipsia*. They are believed to have lived
on the sea floor in sheltered parts of coral reefs. Phillipsiidae was
the last family of trilobites to be represented in the fossil record,
finally disappearing in the late Permian period. Related species
occur in Europe and the East Indies.

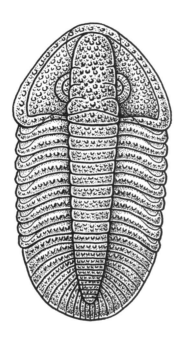

Griffithides bufo Meek and Worthen

Geologic range: Mississippian

Order: Proetida

Family: Phillipsiidae

Where found: Indiana (Keokuk)

COMMENTS: Related species in Maryland, Oklahoma, Texas, Iowa, East Indies, and Europe (e.g. Ireland). This genus is closely related to *Breviphillipsia* and is also a member of the last family of trilobites to become extinct.

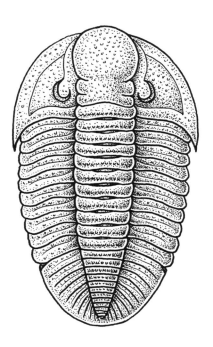

Ameura major Shumard

Geologic range: Pennsylvanian

Order: Proetida

Family: Phillipsiidae

Where found: Colorado, Kansas, Missouri, Montana, Nebraska, New Mexico

COMMENTS:
The only Carboniferous (Miss. – Penn.) American trilobite genus without ornamentation and in which the glabella is widest between the eyes. This species is found in shallow water limestones and shales.

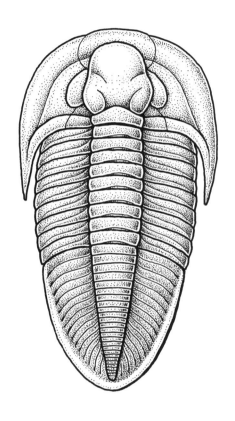

Index

Other Books by Jasper Burns

Johns Hopkins Univ. Press
ISBN 0-8018-4145-3
201 pps. / 450 illustrations
Available in bookstores

Virginia Mus. of Natural History
ISBN 1-884549-03-9
48 pps. / 200 illustrations
Available through VMNH

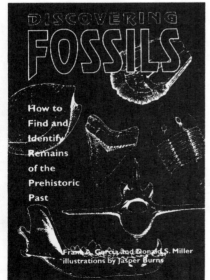

Frank A. Garcia and Donald S. Miller, Jasper Burns (illus.)
Stackpole Books - ISBN 0-8117-2800-5
212 pps. / 170 illus.
Available in bookstores or through Miller's Fossils